CYCLA

A GARDENER'S GUID

by DORIS E. SAUNDERS

Revised and Amended by
R. D. Meikle, B.A., LL.B., (Kew)
C. Grey-Wilson, B.Sc.(Hort.), (Kew)

Editor:
R. C. Elliott, F.L.S.

The Editor acknowledges his indebtedness to Keith Harris, Miss A. V. Brooks, Eliot Hodgkin, Mr. & Mrs. Gerard Parker, Patrick Synge, C. D. Brickell and others for revisions, corrections and additions to the text.

First published in 1959 in the Alpine Garden Society's Bulletin. Revised in January 1973, with further revisions in March 1975.

First reprint, July 1976.

Second reprint, August 1979.

ISBN 0 900048 17 4

PREFACE

The genus *Cyclamen* was reviewed by the late Mrs. Doris Saunders in the Alpine Garden Society *Bulletin* in 1959 (Vol. 27:18). While a very substantial portion of Mrs. Saunders' excellent text remains unchanged and forms the basis of the present publication it was felt that certain alterations and additions must be made to take account of recent nomenclatural and taxonomic changes, and to bring the work as a whole into line with present-day concepts of the genus.

Under current Rules of Botanical Nomenclature, it is our opinion that the following names be adopted, though growers will doubtless resent *C. purpurascens* (for *C. europaeum*) and *C. hederifolium* (for *C. neapolitanum*). *C. coum* now covers *CC. vernum* and *orbiculatum*, but for the most part the nomenclature stands unaltered and is likely to remain so.

The most critical taxa undoubtedly belong to the *C. coum* complex. Here some authorities recognise a number of distinct species; others consider these species only a series of minor variants together constituting a single polymorphic aggregate. In this booklet the aggregate concept is followed.

A new key to the species is included bringing together structurally similar groups of species, and the sequence of species in the text has been re-arranged to agree with this.

The index at the end includes all those names which readers are most likely to come across and their synonomy is indicated where necessary. It is not intended to be a complete list of specific epithets.

One of the chief values of the original manuscript was Mrs. Saunders' remarks on cultivation. These remain virtually unaltered. The section on pests and diseases, however, has been entirely re-written by Keith Harris and Miss A. V. Brooks, respectively, of the R.H.S. Wisley staff.

Some of the plates in the original publication have been retained and a number of new ones added so that most of the species at present in cultivation are now illustrated.

R. D. Meikle and C. Grey-Wilson

INTRODUCTION

The genus *Cyclamen* belongs to the Natural Order Primulaceae and, despite some superficial resemblances to *Dodecatheon* and *Soldanella*, has no obvious affinities with other genera in the family. The first really authoritative account of the genus was given by Hildebrand, and his exhaustive monograph—*Die Gattung Cyclamen* —published in 1898 is still a standard work, although some of his conclusions regarding the delimitation of the species have had to be modified in the light of more recent knowledge.

The subject of classification, however, remains controversial and several species, together with their geographical variants require further study in the wild before their exact botanical status can be determined. Detailed research on the taxonomy and cytology of the genus has been published by Prof. O. Schwarz in *Feddes Repertorium*, Vol. 58:234-283 (1955) and *ibid.* 69:73-103 (1964).

BOTANICAL CHARACTERS

THE TUBER

Cyclamen develop from tubers which, in nature, nearly all undergo a period of dormancy. These tubers differ, according to the species, in shape, in the type of their outer covering and in their root system. The shape is usually spherical, almost completely so in *persicum* and *repandum*, more ovoid in *graecum*, but in the majority of species the upper surface of the tuber is more or less flattened or even slightly concave as in *africanum*, *hederifolium* and sometimes *cilicium*. Two species have irregular tubers, that of *purpurascens* being rounded and knobbly, whilst the tuber of *rohlfsianum* resembles a flattened, somewhat distorted potato.

The outer covering of the tuber is brownish or greyish in colour and may be thick and more or less corky (*africanum*, *cyprium*, *purpurascens*, *hederifolium* and *rohlfsianum*), very corky and inclined to split into flaky, vertical strips (*graecum* and *persicum*), or it may be thin and covered with tufts of very fine hairs as in all other species, except *libanoticum* which loses the hairs at maturity and becomes rather corky.

In most species the roots emerge centrally, occasionally acentrally (*cyprium*), from the lower surface of the tuber only, and those of *graecum* are exceptional in being very thick and fleshy. In *africanum* and *purpurascens* roots are produced all over the surface, in *hederifolium* from the upper surface only and in *rohlfsianum* from widely separated points on the sides and base.

The tubers of some species (*purpurascens*, *graecum*) naturally produce from their upper surface one or more long necks, called floral trunks, at the top of which the leaves and flowers develop. Other species only develop this trunk if the tuber is situated abnormally deeply in the ground (*coum*, *persicum*, *pseud-ibericum* and occasionally *hederifolium*).

FOLIAGE

The leaves of cyclamen vary greatly in size and shape and in texture and colouring. Some have rather dingy leaves (most *coum*, *purpurascens*), others leaves which are so handsome (*graecum*, *persicum*, *rohlfsianum*) that these species are worth growing for their foliage alone. In every species there are forms with better leaves than others and these should always be sought.

Size. This varies from under 2 cm. in length and breadth (a form of *cilicium* collected by E. K. Balls, E.K.B. 669a, and some forms of *coum*) to nearly 10 cm. in length and over 7 cm. in breadth (*africanum*, some *hederifolium*, some *persicum*, *rohlfsianum*—whose leaves are generally broader than long). Most species show a considerable range in leaf dimension and sometimes, in pot-grown plants, the leaves which develop first attain the largest size.

Shape. In shape leaves may be almost circular (*purpurascens*, *coum*), kidney-shaped (*coum*), rounded with a flattened apex (*alpinum*), spoon-shaped (*cilicium*), cordate and sharply pointed (*cyprium*, *pseud-ibericum*), or with a blunt point (*graecum*, *persicum*), or lance-shaped (some *hederifolium*). The margin is frequently cut into lobes (*creticum*, most *hederifolium*, *repandum*, *rohlfsianum*).

Colouring. The colouring and marking are so variable that even in the same species it is almost impossible to find two leaves that are exactly alike. The ground colour is generally some shade of green or grey which ranges respectively from bright green to almost black and from light blue-grey to very dark grey. Occasionally leaves may be almost entirely washed over with silver. Many species have foliage in nearly all these shades, but the leaves of *libanoticum* show little variation in their greyish tone and those of *coum* are nearly always a dark sombre green verging on black. An olive tone is generally characteristic of the leaves of *cyprium*.

The majority of leaves bear on their upper surface a more or less conspicuous spear-shaped or obcordate pattern. It runs from the base towards the apex and is sometimes entire but is more frequently cut into sharply pointed lobes. This "hastate pattern" is generally darker than the rest of the leaf, but may sometimes be lighter or contain lighter areas or marblings within the main pattern. It is surrounded by a broad pale, or silver, band which may be continuous or be separated into variously shaped blotches which in some cases are reduced to mere spots. The brilliance and shape of the patterning vary considerably among individual plants, and some species show a great variety of beautiful and unorthodox markings (*graecum*, *persicum*). In a few cases plain unmarked leaves sometimes occur (*coum*, *purpurascens*, *rohlfsianum*). There seems to be no rule concerning the colour of the under surface, most species exhibiting impartially foliage with either a crimson or a green reverse or even a mixture of the two colours. But there is a tendency in certain species for one or other colour to predominate. Thus the leaves of *hederifolium* and *graecum* are usually green on the back and those of *balearicum*, *cyprium* and *pseud-ibericum* crimson. In some cases colour may change with the age of the plant.

Development. In most species the petiole creeps horizontally under the soil for some distance before assuming a vertical position. Exceptions are *africanum*, *cyprium*, *graecum* and *persicum*, in which the petioles grow almost straight upwards from the tubers. In a few species (*balearicum*, *libanoticum* and *repandum*) the leaf blade emerges from the earth folded tightly inwards along the midrib and does not flatten out until the leaf is almost fully grown. In most other species the young leaves are only partially folded for quite a short period during their development.

THE FLOWER

The general appearance of a cyclamen flower is well known owing to the popularity of the florists' forms of the Persian Cyclamen with their very large, highly coloured flowers. The flowers of the wild species are much smaller than these but possess greater delicacy and charm. They are long lasting and are produced over a long period.

The species may be roughly grouped into autumn, winter and spring flowering kinds, but, since in each group there are usually individual plants or races which come into flower earlier or later than the majority, the seasons tend to merge into one another with the result that cyclamen flowers may be found at almost any time of the year. The greatest floral display takes place during the late autumn and early spring months, a time when colour is especially welcome. The slack season is June and July when all the deciduous kinds are at rest and only the evergreen *purpurascens* is left to carry on.

Size and shape. Since the size of the flower varies from 1.5 cm. (E.K.B. 669a) up to 4 cm. (*africanum*, *libanoticum*, *persicum*), the range may be described as relatively wide. Variation in shape is less marked. The flowers of all species are pendent, so that the mouth of the flower points downwards. The petals are usually reflexed through an angle of 90 degrees, sometimes 180 degrees, and are partially twisted or rolled longitudinally outwards, i.e. away from the calyx. This tends to make them look narrower than they really are, but is an attractive feature. A few species have small ear-like appendages, known as auricles, on the rim of the corolla tube. These are most prominent in *africanum*, *cyprium* and *hederifolium* and slightly less so in *graecum*, although in the island forms of this species from Crete and Cyprus the auricles are quite as pronounced as in *hederifolium*. There is also a form of *purpurascens* (*C. colchicum* Alboff) which is minutely eared. The shape of the corolla tubes and that of the beautiful leaf-like calyces also vary in the different species, but these details are of specialised rather than general interest and may be studied in Hildebrand's monograph on the genus.

Colour. Pink and carmine in a variety of shades are the commonest colours of the flowers, and those species whose flowers are naturally white (*balearicum*, *creticum*, *cyprium*) may sometimes have pale pink or pink-flushed flowers, or be pink on first opening. Many of the coloured species throw albino forms, although in a few cases these are only known horticulturally (*hederifolium*). The white form of *repandum*, long known in cultivation, was only found in the wild in 1949. It is probable that in course of time wild albinos of all the species will be discovered. The habitats of some species are so distant or far from the beaten track that few observers have the opportunity of studying the plants in the mass throughout their geographical range, and even among species which can easily be seen by the thousand white forms are exceedingly rare. When once discovered, the albinos usually breed true if self-fertilised or crossed with each other.

The flowers of most species bear at the base of each petal, near the rim of the mouth, a distinctive red or purple blotch which varies in shape and is characteristic for each species. Albino forms may either carry this marking or be entirely white. It is absent in *balearicum*, *creticum* and E.K.B. 669a, but the mouth of the last named is closely veined with light grey, giving it a greyish tinge. This grey colouration is not known in any other cyclamen.

Many growers of cyclamen have observed their curious habit of producing, from time to time, flowers of a different shade or colour from that of the previous season. The change may be of degree only but in some instances flowers which were white one year may be pink the next, or vice versa. The colour change can take place equally in plants which have been re-potted and in those remaining in the same compost. This peculiarity was well known to Hildebrand and is described in his monograph.

Scent. Few accounts of cyclamen mention scent in connection with more than three or four species yet, actually, eleven species and one variety are fragrant. Of these, five, namely *balearicum*, *creticum*, *purpurascens*, *persicum* and some forms of *rohlfsianum*, have the fragrance of Lily of the Valley; two, *cilicium* and *alpinum*, the unique smell of heather honey; *cyprium* has an acrid spicy scent, and the remaining four, some forms of *africanum* and *hederifolium*, *pseud-ibericum* and *repandum*, have an elusive scent of violets which seems to depend for its intensity on suitable atmospheric conditions.

THE FRUIT

The seeds of cyclamen are borne in spherical or ovoid, greyish-brown capsules which vary in size according to the species, *graecum* and *persicum* having the largest capsules—and seed—whilst *balearicum* has the smallest. After fertilisation of the flower, the corolla falls off and in every species except *persicum* the flower stem, or peduncle, starts to coil up. Again with one exception (*graecum*) the coiling begins at the apex and continues until the developing capsule is carried down to ground level completely enveloped in the coils. In *graecum*, however, coiling begins at the base and proceeds upwards but usually ceases before the capsule is reached, leaving it sticking out horizontally from the bunch of coils.* In *persicum* there is no coiling at all but the peduncle gradually bends over sideways until the capsule is finally deposited on the earth at some distance from the parent plant.

In cultivation the seeds of nearly all the species ripen in July or August. Thus seed of autumn-flowering cyclamen such as *cilicium*, *graecum*, *hederifolium*, etc. take almost ten months to mature, that of the winter flowerers about six months and those that flower in spring only three or four. *C. purpurascens* is particularly slow in ripening its seed and may take more than twelve months according to whether the previous season's flowering period was in early or late summer.

Fresh seed is light golden or chestnut brown in colour and is embedded inside the capsule in a sticky, gelatinous substance which is attractive to insects such as wasps and ants. These will eat the jelly either *in situ*, pushing aside each seed in turn after its sticky coating has been consumed, or carry the seeds to some considerable distance before beginning their feast. This aids dispersal and accounts for the curious places in which seedlings sometimes appear. Seeds are also naturally removed from the immediate vicinity of the parent tuber by the slackening and unwinding of the peduncles which occurs shortly before the capsules are ready to split open.

CYTOLOGY

Glasau investigated the cytology of *Cyclamen* in order to determine the chromosome numbers of the different species, and the possible relationship between chromosome number and morphological characteristics. Later continuing the work of De Haan, Doorenbos and others, O. Schwarz and L. Lepper [*Feddes Repertorium*, Vol. 69 : 73-103 (1964)] have recently investigated and revised the cytology. Investigations show that the numbers range in a very irregular sequence from 20 large to 96 very small chromosomes and to still higher figures in the case of the tetraploids of *graecum* and *persicum*. It was found that, while a few species with identical chromosome numbers bear a certain resemblance to each other, many others belonging to the same chromosome group are widely dissimilar in general appearance, although they may have some morphological similarity, such as the shape of the tuber

* This peculiarity was first noted by Hildebrand (*Berichte Deutsch. Bot. Gesellsch.* Vol. 24:559—562) in 1906, but seems to have been overlooked by other authors, until rediscovered by Dr. N. Mutch and Mrs. Saunders. (See *Curtis's Botanical Magazine* N.S. 603 (1971). R.D.M. also *Proc. R.H.S.* 1959 pp. 53-4 Min. Scientific Committee).

and the type of root system. On the other hand, some species which exhibit physical similarities are dissimilar in chromosome number.

For convenience of reference, the following table is quoted from Schwarz & Lepper.

NUMBER OF CHROMOSOMES		SPECIES
20 large	..	*C. balearicum, C. repandum.*
22 ,,	..	*C. creticum.*
30 ,,	..	*C. cilicium, C. cyprium, C. libanoticum, C. coum, C. pseud-ibericum.*
34 medium	..	*C. africanum* Boiss. & Reut. em. Schwz. & Lepper, *C. hederifolium, C. purpurascens.*
48 small	..	*C. persicum.*
68 medium	..	*C. africanum* auct. (*C. commutatum* Schwz. & Lepper)
84 small	..	*C. graecum.*
96 small	..	*C. rohlfsianum.*

DISTRIBUTION & HABITAT

Introduction. Cyclamen belong entirely to the Old World and, as Mountfort states (*Bull. Alp. Gard. Soc.* Vol. 14, pp. 8-25), "the Mediterranean area is their focus of distribution and they are mainly grouped from its centre to the east". Taking the tip of Italy roughly as the centre, only five species, *africanum*, *balearicum*, *purpurascens*, *hederifolium* and *repandum*, occur west of this point. Of these, only the first two belong exclusively to the western sector, inhabiting respectively Algeria and the Balearic Isles. The remaining three species are also found to the east of the centre but do not extend beyond the coast of Asia Minor except for an outlying form of *purpurascens* (*ponticum*) near the Black Sea. The species of the eastern sector do not overlap into the western sector, apart from a form of *persicum* (*punicum* Pomel) which occurs in Tunisia, but they extend as far east as the Caucasus and N.W. Iran. It is of interest to note that Spain, and possibly Egypt, are the only countries with any Mediterranean coastline which contain no cyclamen.

Most cyclamen are shade lovers and generally grow in positions which are well drained, such as the clefts of rocks, stony banks or walls or the wooded slopes of hills. The woodland species are found, often in incredible numbers, in beech, chestnut and oak woods and in coniferous forest, also in scrub and under bushes, usually in soil containing numerous stones. A few grow in stone slides and under stone slabs, sometimes in full sun.

Although some cyclamen are found near sea level, the majority are inhabitants of hills and mountains generally between the 300 m.

and 1250 m. contours, though some alpine species may reach 2150 m. Most species have a fairly wide altitudinal range.

The depth at which the tubers grow varies from just below the surface to about 30 cm., even among the same species, but the average depth seems to be between 8 and 16 cm.

Cyclamen inhabiting the hotter zones of their distribution get little or no rain for several months in the summer, during which time they become entirely dormant and generally remain so until the onset of the autumn rains. Species from areas where snow lies all the winter are efficiently protected from severe cold by the snow cover and will not reappear until it melts in the spring.

KEY TO THE GENUS CYCLAMEN

1 Anthers exserted in a cone from mouth of corolla tube; leaves kidney-shaped, often deeply lobed (1) C. ROHLFSIANUM (Libya)

1 Anthers included in corolla tube; leaves heart-shaped or rounded—2

2 Corolla lobes auriculate at base; flowers appearing in late summer and autumn, usually before the leaves—3

2 Corolla lobes not auriculate at base—6

3 Fruiting pedicel coiling from the middle or the base; tuber rooting only from the base, with thickened retractile roots; leaves with a thickened cartilaginous border, generally without distinct lobes or angles (2) C. GRAECUM (Aegean Region & Cyprus)

3 Fruiting pedicel coiling from the apex; tuber usually rooting from the upper surface, without thickened retractile roots; leaves without a thickened cartilaginous margin, generally lobed or distinctly angled—4

4 Calyx lobes narrowly triangular, subulate, with a long acuminate apex (5) C. CYPRIUM (Cyprus)

4 Calyx lobes broadly triangular, almost as broad as long, with an abrupt cuspidate apex—5

5 Leaves up to 10 cm. wide; flowers 1.5—2.3 cm. long, mostly appearing in advance of the leaves (3) C. HEDERIFOLIUM (S. Europe & W. Turkey)

5 Leaves up to 20 cm. wide; flowers 2.5—3.5 cm. long, mostly appearing at the same time as the leaves (4) C. AFRICANUM (Algeria)

6 Fruiting pedicels not coiling; flowers large, appearing with the leaves in winter and spring
(6) C. PERSICUM (E. Mediterranean)

6 Fruiting pedicels coiling—7

7 Flowers appearing in summer and autumn—8

7 Flowers appearing in winter or spring—10

8 Flowers deep carmine pink or purple, strongly scented; mouth of corolla 8—10 mm. diameter; corolla lobes broad and blunt; leaves persistent (7) C. PURPURASCENS (Central Europe eastwards to Caucasus)

8 Flowers white or pale pink with a distinct basal blotch, not strongly scented; mouth of corolla usually less than 5 mm. diameter; corolla lobes narrow and acute; leaves not persistent—9

9 Tuber large with a rough corky surface; flowers generally appearing with the leaves; corolla lobes glandular on inner surface; leaf marbling often pink or reddish, at least on the young leaves (9) C. MIRABILE (S.W. Turkey)

9 Tuber small, velvety; flowers generally appearing in advance of the leaves; corolla lobes eglandular or very sparsely glandular on the inner surface; leaf marbling white or greyish
(8) C. CILICIUM (S. Turkey)

10 Flowers white or veined pale pink, without a dark basal zone or basal blotches—11

10 Flowers pink or magenta or with a darker basal zone or basal blotches—13

11 Style included in the corolla tube or very shortly exserted—12

11 Style distinctly exserted from mouth of the corolla tube
(10a) C. REPANDUM ALBUM

12 Corolla veined pale pink; lobes 10—18 mm. long.
(12) C. BALEARICUM (S. France and Balearic Islands)

12 Corolla white without pink veins; lobes 16—25 mm. long
(11) C. CRETICUM (Crete)

13 Corolla with a dark basal zone but without blotches; leaves lobed or angled—14

13 Corolla with distinct basal blotches; leaves not lobed or angled—15

14 Corolla bright purple-pink (10) C. REPANDUM
(N. Mediterranean from France to the Aegean)

14 Corolla white with a pink or purple basal zone
(10b) C. REPANDUM var. RHODENSE (Rhodes)

15 Corolla mouth 10—13 mm. in diameter; corolla pale pink with dark basal blotches (13) C. LIBANOTICUM (Lebanon)

15 Corolla mouth less than 10 mm. in diameter; corolla magenta (white in albino forms)—16

16 Basal blotches without a white or pinkish eye—17
16 Basal blotches with a white or pinkish eye—18

17 Corolla lobes spreading, 10 mm. or more in length (16) C. ALPINUM (S.W. Turkey)

17 Corolla lobes reflexed, generally less than 10 mm. in length (17) C. PARVIFLORUM (N.E. Turkey & Caucasus)

18 Tuber with a rough corky surface; leaves often strongly toothed; corolla lobes with a large and prominent basal eye (14) C. PSEUD-IBERICUM (S.E. Turkey)

18 Tuber smooth, velvety; leaves entire or weakly toothed; corolla lobes with a small basal eye—19

19 Leaves kidney-shaped or sub-orbicular; corolla lobes blunt or rounded, usually not much exceeding 10 mm. in length, with a white basal eye (15a) C. COUM ssp. COUM (E. Europe and W. Asia, including Caucasus and S. to the Lebanon)

19 Leaves heart-shaped, acute, or sub-acute; corolla lobes acute, sometimes exceeding 15 mm. in length with a pink or pale purple eye (15b) C. COUM ssp. CAUCASICUM (N. Iran and Caucasus)

Attention is drawn to the Index on page 49, which not only allows the reader to locate each species mentioned in the text, but is also a guide to synonyms.

THE SPECIES

1. CYCLAMEN ROHLFSIANUM Aschers. (Illustration p. 19).

Colour plate *Bot. Mag.* N.S. t. 192

C. rohlfsianum is a native of Libya, with a very limited distribution along the green belt of Cyrenaica from Benghazi to Derna, where it grows in gullies and on the lower slopes of hills, often in limestone outcrops.

The tuber globular at first, but later may become flattened, with a concavity below, and is often distorted in appearance. The light brown skin, which is very corky, has a curious habit of sloughing off in patches giving the tuber a ragged look and providing points of entry for soil pests and fungi. Any such skinless areas should be painted with methylated spirits which quickly dries them and causes them to callus over. Should rot set in, the diseased portions can be safely cut out without harming the plant. The roots emerge from various parts of the tuber, occasionally from the upper surface but usually from the sides and base. They are pale chestnut-brown and rather shaggy.

The foliage grows from widely scattered points on the top and sides of the tuber and develops only very slowly. The young petioles and unfurled leaves are covered with pinkish glandular hairs which disappear as the foliage matures. The leaf, which is borne on a coral-pink petiole, is broader than long and has a diameter of over 10 cm. when fully developed. It is thus unique in appearance, with a finely dentate margin cut into a number of rather deep, somewhat angular lobes which are sub-divided into shallow lobes with blunt points. In the best forms the upper surface of the leaf is bright shining green handsomely patterned with silver, with the undersurface beetroot red. Other forms may have plain, mid-, or dark-green leaves with a green reverse. The leaves are rather thin and are very easily damaged by frost, which causes unsightly transparent patches, followed rapidly by the death of the leaf.

The flower is also unique by reason of its long exserted stamens and even longer style which project in a cone from the corolla tube, a feature not found in any other *Cyclamen* species, but which occurs in the allied genus *Dodecatheon*. The flower, which is rather larger than that of *hederifolium* and of the same shade of pink, has long lanceolate, sharply pointed petals which are slightly folded and twisted. There are small auricles on the mouth, and the base of each petal bears an upward pointing, broad, triangular, crimson patch which is sometimes bifid at the apex. Some strains of *C. rohlfsianum* are very fragrant and fill the greenhouse with the scent of Lily of the Valley.

Seed is produced freely in cultivation and germinates well, but the young tubers are sometimes difficult to bring through their first year. The seedling leaf resembles the adult in general shape, although without the lobing, and has the same characteristic dentation.

C. rohlfsianum is an autumn flowering species, the flowers generally appearing just before or with the foliage. Where this cyclamen has to be grown without artificial heat it is essential to start it into growth early in August so that the leaves may have at least two months of warm weather in which to develop fully. Leaves which unfold during the colder weather of late autumn rarely attain their full size but remain small and stunted. Only very little water, from below, should be given until the tuber starts to make new growth, otherwise it may rot.

2. CYCLAMEN GRAECUM Link (Illustration p. 19).

Colour plate *Bot. Mag.* N.S. t. 603

C. graecum is a Mediterranean species of wide distribution which is common on the mainland of Greece and on many of the islands, including Crete whose local form was described by Hildebrand as *pseudo-graecum*. It also occurs in S.W. Turkey and in 1948 or 1949 the plant was found in Cyprus. As might be expected from their geographical isolation, the island forms differ in some particulars from the type, the most distinct being the form from Cyprus. Another interesting form is one which was obtained by Glasau from an undesignated island in the Aegean. This plant proved to be a tetraploid to which Glasau in 1939 gave the name "*gaidurowryssii* var. *malingeri*"—both names referring to localities on Mt. Parnes. Since no description of the plant was published, it must be regarded as a *nomen nudum* of no botanical standing. Some writers have regarded such variants as distinct species, and have given them (often invalidly published) names such as *C. cypro-graecum*, but in the light of present knowledge these cannot be looked upon as any more than forms of *C. graecum*.

C. graecum grows in woodland or under bushes, or in the open on stony slopes and beneath stone slabs. The tuber is generally deep seated but may be quite superficial when growing in the clefts of rocks.

The tuber is globular and somewhat flattened and differs from that of all other species by reason of its corky covering which splits into longitudinal strips, and by its long fleshy roots which emerge in bunches from its base. Collected plants often have floral trunks several inches in length which make them difficult to accommodate in a pot. These trunks, however, can safely be cut off to encourage foliage to develop from the surface of the tuber, while the severed trunks may be used as cuttings. If cut off in this way, however, the tuber will take longer to establish.

The foliage of *C. graecum* appears in late August or September just after, or sometimes with, the flowers. It is so variable that it is difficult to differentiate a type form, but what might be described as the ordinary *graecum* leaf is obcordate and pointed, and rich deep velvety-green with a broad, almost continuous band of silver half an inch inside the border of the leaf. The margin is edged with reddish-brown horny teeth. The leaf veins are also silver and very conspicuous. The under-surface is light green. This is, perhaps, the handsomest type of foliage. The leaves of the variant from Crete are inclined to be greyish-green, some being almost entirely washed over with grey, while the band is pale, but not silvery, and is discontinuous, giving the foliage a distinctly chequered look. The under-surface is generally dark beetroot red or green.

The flower of the "ordinary" *graecum* is much reflexed, with a span of half an inch, and the petals are close together and not twisted. Their colour is deep reddish pink or pale rose and the rather small auricles are bluish pink. At the base of each petal there are two dark plum-red blotches separated from each other by a much smaller spot which is not contiguous. From the larger blotches two branched veins of colour run up almost to the tip of the petal, while from the small spot there extends a single vein. The anthers are violet and the stigma pale pink and not exserted.

After fertilisation, the peduncle, unlike that of any other species, starts to coil from the ground upwards, finally leaving the capsule, which is very large, protruding horizontally.

C. graecum has the reputation of being a shy, or non-flowerer incultivation but this is probably because it does not get sufficiently ripened during the summer. The tuber needs all the heat it can get while it is resting and must not be given a drop of water until new growth starts in late August or early September. It is best grown in a deep pot for there are few gardens in which it will flower freely outside; one such is Highdown, where a colony has been successfully established on the limestone cliff: also in the bulb frame at Wisley, where it flowers regularly and profusely.

3. C. HEDERIFOLIUM Willd. (= NEAPOLITANUM Tenore) (Illustration p. 20).

This is the best known and most accommodating of all the species and can be grown successfully in almost every garden where the soil is not too stiff.

The plant is widely distributed along the Mediterranean basin from southern France (doubtfully) to western Turkey where Peter Davis speaks of finding enormous tubers among the ruins of the Greek theatre at Ephesus. It is very common in Corsica and Sardinia, and is found in many of the other Mediterranean islands, including Crete, but does not seem to grow in Cyprus. The island forms are often very fine, one collected on Andros by Davis being particularly handsome and scented, while the race from Corfu has long been famous for its scent. Generally the scent is rather faint, but strongly scented forms are known and the character is sometimes transmitted. This is a very long-lived species, up to 100 years or perhaps more, and E. A. Bowles reported in 1948 on a plant which had been established in 1856.

The tuber is rounded below and slightly flattened above with roots emerging only from the top and sides, a peculiarity that serves to distinguish it from all other cyclamen. It has a dark greyish-brown, corky skin and may often attain a very large size.

The leaves, which appear after the flowers (in the island forms often with them), are produced from the centre of the tuber and are borne on long creeping petioles. They are extremely variable in shape, some being almost rounded, others almost as broad as they are long, others, again lance-shaped. The blades may be entire or cut into shallow lobes with blunt or sharp points. The margins usually bear coarse or fine teeth but are occasionally plain. The colour of the leaves ranges from deep green to light grey, or even silver. The darker leaves are inclined to be sombre, but those of a lighter shade are generally beautifully marbled in a bewildering variety of patterns. The underside is plain green or, more rarely, beetroot red. The foliage is very hardy and stands up to the vagaries of the English winter better than that of any other species. It persists from September until the end of May, after which the tuber rests until late July or August when the first flowers appear.

During the resting period a good top dressing of leaf mould will help to feed the tuber and encourage strong new growth.

The flowers are large and well shaped with long ovate-lanceolate, sharply pointed petals which are constricted near the base and then flare out to form a wide rim to the mouth which has conspicuous auricles. The petals are rose-pink in shades of varying intensity, and at the base of each is a dark crimson blotch from which two narrow prongs of colour extend a short way towards the tip of the petal. The mouth of the corolla tube is pentagonal and the stigma does not protrude. The anthers are brownish-red. A beautiful albino form, with the mouth either pure white or faintly flushed with pink, is well known in cultivation, but almost unknown in the wild. It has, nevertheless, been seen in Greece recently by one of our Members.

C. hederifolium usually does better and looks more attractive in the garden than in a pot, where the creeping habit of the leaf and flower stalks is inclined to produce a sprawling effect. A mulch of sifted leafmould in August in beneficial. The plant does not resent being lifted, and therefore, if required for exhibition, can safely be dug up from the open and can be replanted later. E. A. Bowles paid this fine plant the compliment of saying "it pays rent for 11 months of the year".

4. C. AFRICANUM Boiss. et Reut.

C. africanum is a native of Algeria and comes so close to *C. hederifolium* that some botanists have hesitated to accord it specific rank. However, since Glasau discovered that it possesses 68 chromosomes, twice the number of *hederifolium*, Doorenbos taking into account its other divergencies from that species, has advanced the theory that it may be a tetraploid, not of *hederifolium* itself but of a closely related African species, also from Algeria, which Glasau described as *C. numidicum*. This seems to be almost identical with *africanum*, but has half the number of chromosomes. Unfortunately this interesting plant appears to be no longer in cultivation.*

The tuber of *africanum* is much flattened and slightly concave above and is covered with a greyish-brown, flaking, corky skin. The roots develop over the whole surface of the tuber, unlike those of *hederifolium* which generally spring from the upper surface only.

The leaves, which are borne on ascending petioles, not creeping as in *hederifolium*, start to unfold just before, or with, the first flowers and are not usually fully developed until all the flowers have opened. They are large, but not grossly so, provided the plant is not overpotted nor overfed. They may be entire and almost circular, or reniform, or cordate and pointed, or cordate and shallowly lobed. They are leathery and generally the colouring is a rather dark green with a glazed surface and very little marbling. However, forms with more emphatic tones occur and sometimes the ground colour is dark grey instead of green. The leaf margin is edged with coarse, horny teeth; the undersurface is light green; the leaves are, usually, the largest of any species in the genus.

The flowers have a faint scent of violets and are very handsome, with long, dark rose-pink, sometimes pale pink, petals and well developed auricles. At the base of each petal is a wide crimson patch which is bifid at the tip. The stigma is equal in length to the corolla tube, and the anthers are yellow with a purple groove on the back, in contra-distinction to those of *hederifolium* which are brownish-red. A typical *africanum* is easily distinguished from *hederifolium*, but quite a number of intermediate forms occur in cultivation which may have risen from interbreeding between the two species.

C. africanum is too tender† to be grown outside in most gardens, but does well in a pot provided it is thoroughly dried off during the summer and is not watered until the first flower buds are well above the ground. It should then be given a very little water and thereafter kept rather on the dry side.

* Schwarz and Lepper consider typical *C. africanum* to be the diploid (2n=34) and have described the more commonly cultivated tetraploid (2n=68) as a distinct species (*C. commutatum* Schwarz and Lepper). It is difficult to see how *C. africanum* sensu Schwarz and Lepper can be satisfactorily distinguished from forms of the variable *C. hederifolium*.

† C. D. Brickell reports that it does well in the bulb frame at Wisley.

5. C. CYPRIUM Schott & Kotschy (Illustration p. 22).

This autumn-flowering species appears to be endemic to Cyprus; records from Cilicia, and other areas, are all very questionable.

The plant is very common in certain parts of Cyprus, particularly in the Kyrenia range, growing on woodland banks, or in stony places at altitudes of 300—1200 m. It is also found growing in the shade of cedar forest in the Paphos area. The tuber is round, and is covered with a rough greyish skin and is frequently flattened both above and below. The roots grow in a bunch from the lower surface only, often acentrally, a feature which is apparently peculiar to this species. During the resting period the tuber sometimes sheds every vestige of last season's growth and becomes so smooth that it is almost impossible to ascertain which are its dorsal and ventral surfaces. In this event it should be placed on its side and correctly orientated when potted on. It flowers in late autumn or early winter.

The leaves are generally rather small, about 3.5 cm. in length by a little less across, and are very distinctive. Their ground colour is olive green and there are various light green blotches near the margin. The hastate pattern is long and rather narrow and is edged with a broad irregular band of light green. The underside is bright carmine. In shape the leaf is broadly cordate and sharply pointed, or ovate-lanceolate tapering to a long point, whilst the margin is cut into very shallow lobes with sharp tips terminated by pale yellow knobs. Unblotched leaf forms also exist.

The flower, which is very fragrant, is much reflexed, with the petals folded near the base and twisted towards the apex. In a good form the flower is long and graceful and can exceed 2.5 cm. in length. The petals are pure white when fully developed but sometimes open shell pink. Their tips are often irregularly and distinctly toothed. The mouth bears conspicuous auricles and at the base of each petal is a broad V-shaped pinkish-purple blotch of a shade not found in any other species. The corolla tube is narrowly campanulate and not constricted; the style is not exserted.

C. cyprium grows well and flowers freely in a pot, but must be protected from severe frost. Lewis Palmer reported that it grew better below the staging than on top: he also grew it successfully out-of-doors in his garden near Winchester. A photograph of *cyprium* in the wild, together with an interesting account of the plant, will be found in *Bull. Alp. Gard. Soc.*, Vol. 17, pp. 240-242.

6. C. PERSICUM Mill. (Illustration p. 21).

An inherent tendency to vary is present in all cyclamen species but this characteristic reaches its extreme in *C. persicum*, a fact which was put to good advantage by the plant breeders whose work with the plant eventually produced the enormous range of large-flowered and variously-coloured florists' strains that we know today. The wild plant shows an almost infinite amount of variation in its foliage but the flower tones range only from pure white, through lilac, to rose pink and orchid purple. The flame, red and almost black shades of the florists' strains are absent. However, this is off-set by the delicious perfume which is nearly always lacking in the horticultural varieties.

The main focus of distribution of *C. persicum* is the eastern Mediterranean where it is locally common, in Crete, Rhodes and the Aegean islands, Cyprus, Turkey (Cilicia), Syria, Palestine, Lebanon and finally much further west, in Tunisia, in the form known as *punicum*. The plant grows in thin woodland and among scrub and may also be found in the open, on hillsides or in the cracks of rocks or walls. It has a wide altitudinal range from near sea level to over 1250 m. It is particularly fine growing, mixed with scarlet ranunculus, in limestone crevices on the lower slopes of Mt. Lebanon.

The tuber is rounded and covered with a corky skin somewhat similar to that of *graecum*, but less striated. The roots emerge from the centre and sides of the lower surface and are inclined to be fleshy. In plants grown without artificial heat the leaves start to appear any time from late autumn onwards according to the temperature, and no water should be given until they are clearly visible. Sometimes an early crop of leaves will die off just as the plant is coming into flower and if a second crop is not produced in time to coincide with the full flowering period the appearance of the plant is spoilt.

The foliage is extremely beautiful and variable in colour, marking and size. It also varies considerably in shape, although this is usually cordate, sometimes overlapping at the base, more or less pointed and with an evenly toothed margin, the teeth being horny and closely set. Very occasionally the leaf is lobed. The undersurface is most commonly green, but forms occur with a dark red or mottled reverse.

The flowers appear in early spring and are carried well above the leaves on peduncles up to 20 cm. in length. They are long and graceful with much reflexed, attractively twisted petals. The colour is most often a pale lilac white, but pure white forms occur, and there are very good pale and deep pinks, which unfortunately do not seem inclined to breed true to colour. On each petal there is a conspicuous dark purple rectangular blotch which extends for some distance up the petal before narrowing slightly. The style hardly exceeds the corolla tube. The flowers are extremely fragrant with a Lily of the Valley scent which is so attractive to insects, particularly bees, that sometimes too many flowers are fertilised. If all of these are allowed to set seed the tuber may be seriously weakened and be unable to flower the following season. After fertilisation the peduncles do not coil, as in all the other species, but increase in length and gradually bend down to the ground. Seed germinates well provided it is kept in a damp atmosphere.

C. persicum must be regarded as a half-hardy plant and be adequately protected from frost, otherwise all the foliage will die off even if the tuber is not damaged.

7. C. PURPURASCENS Mill. (=EUROPAEUM L.) (Illustration p. 23).
This species comes from limestone districts as far west as Haute Savoie and Southern Switzerland, but its main distribution is to the east, where it becomes increasingly common. It is found, often in beech woods, in southern Austria, northern Italy and Yugoslavia and extends to Macedonia and Bulgaria. A variety (*ponticum*) occurs in the Caucasus. This form is regarded as a species by Schwarz and is also given specific rank by Pobedimova.

The tuber is rounded, sometimes flattened and rather irregular in outline. Occasionally small excrescences grow out from it, or a new tuber may appear beneath the old one, connected to it by a short trunk perhaps several inches in length. The tuber is covered with a chestnut-brown, sometimes rather corky, skin, and roots are produced from all over the surface. When growing very deeply, as is frequent in the wild, it produces a floral trunk which carries the leaves and flowers up towards the surface of the soil. It should never be dried off.

In nature the plant is usually covered by snow for part of the winter months and thus undergoes an enforced period of dormancy, but the foliage is naturally evergreen. When protected from severe frost and cold winds it will retain most of its old leaves for a month or more after the new crop has appeared. The leaf is almost circular, the margin edged with small regular teeth, or occasionally plain. The ground colour is a rather sombre green, sometimes very dark, relieved by a bold or inconspicuous patterning of white. In some cases the leaves are plain green without any markings. The under-surface is most often dark red but may sometimes be green or a mixture of the two.

The flowers are very sweetly scented and for this reason are extensively picked in the districts where they abound. The colour is rather variable, good forms being rose-pink or carmine with darker veins, poor ones a rather slaty magenta. At the base of each petal is a broadly oval crimson patch. There are normally no auricles, although eared forms are known. The petals are rather short and spreading which gives the flower a somewhat dumpy appearance, but it is possible to find more graceful forms. A very rare albino is known, but is probably no longer in cultivation. Sir Austen Chamberlain found it near Geneva, and there was a small colony which the authoress saw in N. Croatia.

C. purpurascens is a summer-flowering species, starting any time from early June onwards according to locality. In cultivation the flowering period may last until the end of December. When happy, the plant is very free flowering, but is not always easy to suit. It seems to require very good drainage and probably does best in a litter of beech or oak leaf mould mixed with lumps of limestone. *Cyclamen fatrense* Halda et Sojak, which has recently been described from Czechoslovakia, is little more than a pure green-leaved form of *C. purpurascens:* such variation is common throughout the range of this species. *Cyclamen colchicum* Alboff is a species of which little material exists, but it appears to be very close to, if not, *C. purpurascens*. Seedlings have recently been distributed.

8. CYCLAMEN CILICIUM Boiss. et Held. (Illustration p. 22).

This species is largely confined to the province of Cilicia in S.W. Turkey. It occurs in wooded mountainous regions, usually in coniferous forest, at altitudes of over 1250 m.

The tuber is covered with a thin greyish-brown skin and is much flattened, with roots emerging from the centre of the underside.

The leaves appear in late September or October, generally with the flowers. They vary considerably in size and marking but are always spoon-shaped rather than rounded, with a wavy margin and prominent endings to the leaf veins. The general effect is greyish-green, although some plants have very dark leaves. The hastate mark is usually narrow and is surrounded by a band of irregular pale blotches and a few darker spots, giving the leaf a speckled appearance. The better forms have bolder and more definite markings with sometimes a complete ring of light spots round the margin of the leaf. The underside is usually beetroot red.

The flower, although small, has an elegant shape with long sharply pointed petals which are folded slightly outwards at the base. They vary in colour from almost white to deep rose-pink, and have pink veining. There are no auricles and the style is equal in length to the corolla tube. Near the base of each petal is a conspicuous crimson, dome-shaped blotch from which three faint prongs of colour extend towards the mouth, whilst the upper portion of the blotch is prolonged into a number of fine streaks, those in the centre being the longest. The flowers appear in late autumn and smell delightfully of heather honey. A plant with salmon pink flowers was recently collected in Turkey and sent to Holland.

C. cilicium is an excellent pot plant but is seldom happy in the open. (This view is not universally shared, for it grew abundantly both in Lewis Palmer's cold garden near Winchester and in the peat banks in E. B. Anderson's Cotswold garden. At Highdown it thrived in a warm situation in gritty soil. Editor)

I did find out I am not yet sufficiently clear about to put into words.

We all hoped that we should spend this Christmas and usher in the New Year under conditions of peace. Our hopes are deferred, but still we, I and all the staff of MY GARDEN, wish every reader a Happy Christmas and Health and Peace in 1945.

T. A. S.

The Blue Spiræa

It is 100 years since Fortune introduced from China Caryopteris Mastacanthus, popularly called "Blue Spiræa," though it belongs to the verbena family. Averaging about 6 ft., this presents a slender-branched symmetrical shrub, yielding towards the end of its many twigs cymes of violet flowers. The leaves are pleasantly aromatic and the shrub is valuable since it blooms late, even into autumn when such things are few. Though at one time treated as a greenhouse shrub, it is quite hardy enough to endure an average winter in most places. But good as it is, C. Mastacanthus has been out-classed by the hybrid from it and C. mongolica called C. clandonensis. Making a rounded bush much after the style of the foregoing, this is a most beautiful object from August onwards, its prolific crop of whorled blossoms being a very attractive leadwort-blue. This also is hardy, but both shrubs should have a light soil and full sun is desirable.

Cyclamen repandum

(*See opposite page*)

THE hardy cyclamens girdle the year round with their winning beauty, each season being represented by one or more. Thus no sooner have the crimson blossoms of C. ibericum departed with the lengthening days of the early year than C. repandum (vernum) crops up to linger on until those of summer come along with their sweetness and grace.

C. repandum may not appeal to some so strongly as others, for the ardent red of its blossoms may have a suspicion of magenta in their make-up. But, coming as they do when there is little else, these daintily modelled flowers are exceedingly welcome and, like others of its kind, C. repandum is an easy plant to establish. In Italy and Corsica it grows in shady walls, and in walls, or a rock crevice, it is happiest here. All you have to do is to find a convenient hole about the size of a tennis ball, or less, see that it is well packed with soil, and push in the tuber, making it firm with a little more soil. Late summer or early autumn is the best time for this, and the position should be a cool one, or partly shaded.

The leaves of C. repandum appear with the flowers and they are singularly beautiful. Heart-shaped and thin in texture, the margins are prettily goffered, and in colour they are olive-green with a chevron of silver and faint marblings of jade. This cyclamen shares with others a reputation for longevity, improving in size and prolificacy as the seasons pass, and that without asking anything in the way of cultural aid.

CYCLAMEN REPANDUM.

(1) *C. rohlfsianum* (p. 10)

(2) *C. graecum* (p. 11)

Photos: Downward

(3) *C. hederifolium* (*neapolitanum*) (p. 12) *Photo: Downward*

(6) *C. persicum* (p. 15) *Photo: Downward*

(5) *C. cyprium* (p. 15) *Photo: Downward*

(8) *C. cilicium* (p. 18) *Photo: F. W. Buglass*

(7) *C. purpurascens* (*europaeum*) (Julian Alps) (p. 17) — *Photo: M. G. Hodgman*

(8) *C. cilicium* var. (E. K. Balls 669a) (p. 31) *Photo: V. Cohen*

(9) *C. mirabile* (p. 31) *Photo: R. Elliott*

(10) *C. repandum* (p. 32)

Photos: Downward

(10) *C. repandum*, Peloponnese form (p. 32)

(11) *C. creticum* (p. 33) Photo: Downward

(12) *C. balearicum* (p. 34) Photo: Malby

(13) *C. libanoticum* (p. 34) *Photo: Downward*

(14) *C. pseud-ibericum* (p. 35) (Amanus Mts.) *Photo: J. M. Watson*

(15a) *C. coum* ssp. *coum* (p. 37) *Photo: T. C. Clare*

(15a) *C. coum* ssp. *coum* (E. K. Balls 371) (p. 37) *Photo: Downward*

(16) *C. alpinum* (P.D. 25579) (p. 38) — *Photo: Downward*

(17) *C. parviflorum* (p. 39) — *Photo: R. Elliott*

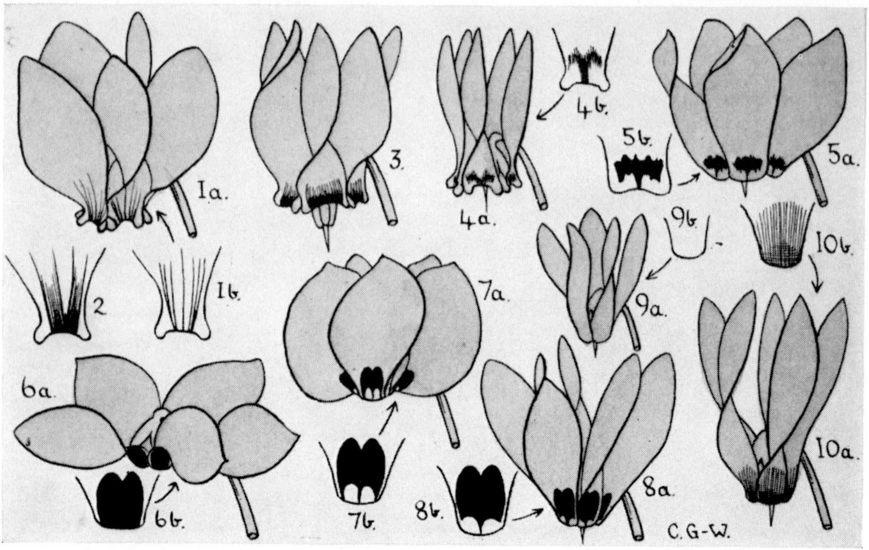

I. SPECIES WITH AURICLES

1a. *C. graecum*
1b. *C. graecum* sinus
2. *C. hederifolium* sinus
3. *C. rohlfsianum*—note anther cone
4a. *C. cyprium*
4b. *C. cyprium* sinus

II. SPECIES WITHOUT AURICLES

5a. *C. libanoticum*
5b. *C. libanoticum* sinus
6a. *C. alpinum*
6b. *C. alpinum* sinus
7a. *C. coum*
7b. *C. coum* sinus
8a. *C. pseud-ibericum*
8b. *C. pseud-ibericum* sinus
9a. *C. creticum*
9b. *C. creticum* sinus
10a. *C. persicum*
10b. *C. persicum* sinus

CYCLAMEN CILICIUM var. (Illustration of E.K.B. 669a p. 24).

Colour plate *Bot. Mag.* N.S. t. 307

This miniature plant, which is the smallest cyclamen known, was collected by E. K. Balls in Asia Minor in 1934. It was growing among limestone rocks under the shade of *Abies cilicica* at 1250 m. at Burujik, close to the famous pass known as the Cilician Gates. The plant was out of flower but because it came within the geographical range of *cilicium* it was assumed to be that species. When, however, it was subsequently studied in cultivation it was found to differ in several respects from *cilicium*.

The tuber is much smaller and at maturity barely exceeds 4 cm. in diameter. The leaves appear earlier, are smaller, differ in outline, being circular or kidney-shaped, and have scalloped edges with the leaf veins ending in sunken pits as in *coum*. In colour they are a curious shade of brownish-green with a broken ring of dull silvery-green blotches surrounding a somewhat ill-defined hastate pattern. The underside is light green. The flowers, which appear earlier than those of *cilicium*, are similar in shape but only about half the size. They are either white or pale pink, the colour varying from one season to the next. The petals are veined, not with pink as in *cilicium* but with light ash-grey. There is no crimson marking at the base, but the mouth looks grey owing to the convergence of the veining. The style is more exserted than that of *cilicium*. The flower has no scent. The plant is perfectly hardy but owing to its small size is better grown in a pot than in the open.

This interesting plant has now been examined at Kew and has been named *C. cilicium* var. but without any varietal epithet. The pink flowered form is described and figured in *Curtis's Botanical Magazine* N.S. t. 307 (1957). In this description the absence of any red colouration on the mouth is noted, but otherwise no attention is drawn to the considerable differences between Balls' plant and typical *cilicium*.

The descriptive epithet "*alpinum*" which was attached to this plant on its introduction, presumably to distinguish it from true *cilicium*, is still in common use, but should not be confused with *C. alpinum* Spreng. As this has no botanical standing it is unfortunate that no better designation than "var." can be legitimately applied.

9. CYCLAMEN MIRABILE Hildebr. (Illustration p. 24).

Colour Plate *Bot. Mag.* N.S. t. 579

This species was described by Hildebrand in 1906 from a single cultivated plant. It was named *C. mirabile* because the young leaves were zoned bright carmine pink above, and not silvery-white as in most cyclamen. The plant examined by Hildebrand failed to survive and *C. mirabile* was lost to cultivation until Davis and Polunin collected another specimen in W. Turkey on the 23rd March 1956. Nine years later additional material was collected in this same area and since then the species has been introduced in considerable quantity so that it is now fairly well established in cultivation.

Superficially *C. mirabile* bears a very close resemblance to *C. cilicium* and flowers at about the same time of year. It has a relatively large corky, somewhat fissured, tuber 3—6 cm. in diameter and sub-orbicular or reniform leaves, usually rounded or obtuse at the apex, about 1.5—3.5 cm. wide, with the margin distinctly but rather remotely toothed. The upper surface of the leaf is dull green with a median zone of pink or silvery-grey marbling. The peculiarity of the marbling which induced Hildebrand to give the species its name is now known to be inconstant; some plants are conspicuously pink-zoned, others soon lose their pinkness and some are silvery-marbled from the very earliest stages of their development. The flowers arise directly from the tuber on erect (not geniculate*) peduncles and have pale pink corollas with a small purple basal blotch. Superficially they are very much like those of *C. cilicium* but the corolla lobes are normally rather more conspicuously toothed at the apex and the inner surface of the lobes (the outer once the lobes reflex) are covered with small glands. The calyx lobes are usually broader and more obtuse than those of *C. cilicium*.

C. mirabile is now known to be locally fairly abundant on metamorphic and limestone rocks in the vilayet of Mugla and Isparta and may be expected to occur elsewhere in S. Turkey. For the present it is best regarded as a distinct species but further collections may show that the distinction is difficult to maintain and that the two, *C. mirabile* and *C. cilicium*, are better regarded as segregates of a single variable species.

**Geniculate, abruptly bent. Ed.*

10. C. REPANDUM Sibth. et Smith. (Illustrations p. 25).

This species is widely distributed in the Mediterranean basin, occurring in the south of France (very locally), Corsica, Sardinia, Italy, Sicily, Dalmatia, Greece, Crete and Rhodes. *C. repandum* is essentially a woodland plant and its very thin leaves are easily damaged by exposure to hot sunlight.

The tuber is globular and much flattened, with a hairy, chestnut-brown skin, and roots coming only from the lower surface. The new foliage begins to grow in the late summer and in mild winters the leaves of pot grown plants may appear above the ground before Christmas and will need protection in severe weather.

The leaf is very distinct and not to be confused with that of any other cyclamen. It is broadly cordate, tapering quickly to a sharp point, and the basal lobes do not overlap. The margin is sharply toothed and shallowly cut into wide lobes each ending in a spiny point. The colour varies from mid to very dark green, with the hastate pattern broadly outlined in silver. An attractive form is one with very dark (sometimes almost black) leaves brilliantly patterned with white. The form from the Greek Peloponnese, which is also illustrated, has large dark green leaves speckled and dashed with cream. The underside of the foliage in most cases is bright, reddish-purple. The leaf stems creep for some distance under the ground before emerging on the surface.

The flowers, which smell faintly of violets (although there are scented and unscented forms) appear in April and May and are long and elegant with slightly twisted translucent petals of carmine red, rarely pale pink or white. There is a long exserted style. The mouth is without auricles and is coloured dark red.

The plant does well in a pot and is worth trying outside in a sheltered place. In some woodland gardens in the warmer parts of Great Britain *C. repandum* spreads as freely as *hederifolium*.

10a. C. REPANDUM ALBUM the white form has long been known in cultivation, though rare and has recently been found growing wild in Corsica. The flowers are pure white.

10b. C. REPANDUM var. RHODENSE Meikle. This is a distinct plant growing on the Island of Rhodes, with pure white petals and a delicate rose-pink mouth. It is more or less hardy out-of-doors, but should be given a sheltered spot in partial shade. It tends to flower later than other forms. The leaf marks are also strikingly different—these are a greyish-pewter colour speckled with greyish-white. In nature the Rhodes variety is a mountain plant, growing in the forest area of Mt. Profitas Elias, in company with such endemics as *Colchicum macrophyllum*, *Paeonia rhodia* and *Cyclamen persicum*. P. M. Synge reports that, grown in a bulb-frame in his sunny garden, it flowers before *C. repandum* (which he grows successfully outside).

11. C. CRETICUM Hildebr. (Illustration p. 26).

Colour plate *Bot. Mag.* N.S. t. 450

Though perhaps rare in cultivation, it is very abundant in Crete, flowering in early April, and growing in open woodland and shady banks in lower areas.

The tuber is covered with a very thin greyish-brown skin and is extremely flattened, with roots emerging only from the ventral surface.

The leaves appear in autumn and in shape are rather similar to those of *repandum* but are much smaller. They are cordate and sharply pointed and have jagged margins. The upper surface is dark, sometimes very dark, grey, faintly burnished with silver or spotted with light grey. The under-surface is beetroot red.

The flowers, which have a faint Lily of the Valley fragrance, do not usually open until April. They bear a resemblance to those of *balearicum* but are longer and more graceful, with pure white petals, occasionally flushed with pink. The mouth of the corolla does not bear coloured markings.

The plant is easily raised from seed which germinates in autumn. It is slightly more tender than *balearicum* and must be adequately protected from severe frost. It is best regarded as a greenhouse plant.

12. C. BALEARICUM Willk. (Illustration p. 26).

This small species from the Balearic Isles and southern France needs protection during severe weather. The tuber is somewhat flattened, with roots springing only from the centre of the base. The beautiful leaves are narrowly ovate, pointed and shallowly toothed; the upper surface is bluish-grey; marbled with silver or occasionally silver all over; the lower surface is bright crimson. The leaves often appear in late autumn or winter, but the flowers are not produced until the following March or April. They are white, with twisted petals faintly veined with pink, very clumpy, with short petioles. They are very fragrant. The style is hardly, if at all, exserted. In nature the plant is found growing in shady places, often in the clefts of rocks, and on that account, and because of its thin leaves, it should not be exposed to hot sunshine.
(There is a picture of it, taken in the wild, in *Bull. Alp. Gard. Soc.* Vol. 20, p. 24.)

13. C. LIBANOTICUM Hildebr. (Illustration p. 27).
Colour plate *Bot. Mag.* N.S. t. 296

C. libanoticum was first discovered in the Lebanon in 1895. It was named and described by Hildebrand but is not included in his monograph. The plant was originally found on the lower slopes of the Lebanon range, N.E. of Beirut, at an altitude of 750—1400 m. It was growing in shady, rocky positions, sometimes among the roots of trees. At the time of its discovery, and for a considerable period subsequently, it evidently occurred fairly widely in the area, but was subsequently lost sight of, and even thought to be extinct, until re-discovered, in very small quantity, by Père Paul Mouterde in November 1961. The recently found population has in cultivation shown no significant differences from that originally introduced and described by Hildebrand (see E. Hodgkin in *Journ. Royal Hort. Soc.* Vol. 87: 228 (1962)). Schwarz states that *C. libanoticum* occurs in S. Turkey, in the Amanus Mountains, but the plant he refers to this species is apparently *pseud-ibericum*.

The tuber is rounded and somewhat flattened and is covered, when young, by a thin greyish-brown skin which bears tufts of hairs. As the tuber ages, however, these hairs gradually disappear, and the skin becomes thick and rather corky, an interesting transition which has not been observed in other species. The roots are produced principally from the centre of the lower surface of the tuber but a few also emerge acentrally. The tuber is very slow growing and a twenty-year-old specimen may not exceed an inch in diameter.

The leaves, which have creeping petioles, usually appear before Christmas and at first are tightly folded together. When young they are rather similar to those of *cyprium* and seedlings of the two species may be confused. However, the mature leaves are very distinct, particularly in their colour. This is a dull bluish-green with darker blotches, sometimes with a rusty tinge. The hastate pattern is long and narrow and very clearly marked. It may be entirely dark green or may have a lighter central zone. In shape the leaves are rounded or obcordate, occasionally somewhat pointed, and up to 8 cm. in diameter. The margin is slightly wavy and the basal lobes are well rounded and close together, although not usually overlapping. The undersurface is shining beetroot red.

The flower buds usually appear above the soil in December but do not open until February or March. The flowers are very striking, with broad, ovate, sharply pointed petals which show no twisting and very little folding so that they appear very large. On first opening they are white or pale rose pink, but they gradually deepen in colour until the whole flower becomes a beautiful shade of clear brilliant pink. An albino is known in cultivation. The corolla tube is campanulate, with a white rim to the mouth. Each petal bears near its base a light crimson marking shaped like a bird in flight. The anthers are deep yellow, and the style is not exserted. The flowers are long lasting but unfortunately have an unpleasant smell. Unless the weather is very cold, plenty of seed is set and this germinates freely. The seedlings are best left undisturbed for the first two years.

In the warmer parts of England *libanoticum* may succeed out of doors if planted in a very dry, well drained position in full sun (see *Bull. Alp. Gard. Soc.*, Vol. 14, pp. 12, 81 and 89), but as a rule it is safest grown in a pot where, owing to the small size of the tubers, a number should be grown together. These will go on for many years without the need of repotting and will annually give a fine display of flowers.

14. C. PSEUD-IBERICUM Hildebr. (Illustration p. 27).
Colour plate *Bot. Mag.* N.S. t. 417

It is almost sixty years since it was originally described by Hildebrand from a flowering specimen sent to him by the Dutch bulb firm of Van Tubergen. This was part of a consignment collected in Turkey, or possibly near the Syrian border. This form is still in cultivation and is deeper in colour than most of those collected subsequently in Cilicia and in the Amanus Mountains. The honour of the discovery goes to Turkish botanists, but it was Davis and Polunin who re-introduced the species to Great Britain in 1957. It has since been collected by a number of expeditions. The plant appears to be common in the beech woods above Harunye on the slopes of Düldül Dagh and is well known to the local inhabitants. In these same woods grow also *C. coum* and the common primrose. The climate here is a good deal damper than in other parts of Cilicia, which augurs well for the establishment of *pseud-ibericum* in English gardens.

The plant has a globular tuber, somewhat flattened above and covered by a greyish-brown scaly skin, and has roots emerging only from the base. Sometimes a floral trunk is present. The leaves are obcordate and shortly pointed. They are very dark green with lighter marbling and a red reverse. The margin is regularly and sharply toothed. The flowers, which are carried on long peduncles, are brilliant magenta and occasionally lilac, with broadly lanceolate, well spread petals about 2.5 cm. long by 1 cm. broad. There is a broad band of white round the mouth of the corolla tube, and at the base of each petal is a conspicuous chocolate blotch shaped like an ace of spades whose leg bisects the white band, thus giving the mouth a curious split lipped appearance. The flowers are borne in February and March and have a strong fragrance of violets.

The plant is hardy in a pot and would probably succeed in the open under woodland conditions: indeed P. M. Synge reports that it grew well in a dry spot beneath conifers in Lewis Palmer's garden near Winchester.

15. C. COUM Mill. Colour plate (as *C. vernum*) *Bot. Mag.* N.S. t. 29

The Coum Group has a wide area of distribution, stretching discontinuously from the Black Sea coast of Bulgaria, the Pontic coast of Turkey and the area around Istanbul (the only European stations), through Asia Minor to the Caucasus and N.W. Persia. Though named "of Cos" by Miller, there appears no record of it being found on that island. It also extends southward into Syria, Lebanon and northern Israel. In view of the enormous area which this species covers it is not surprising that it should exhibit a number of geographical variants. These forms, some of which show distinctive features, have been classified and named in a variety of ways, according to the lumping or splitting tendencies of the botanists who have undertaken the revision of this difficult group. Doorenbos is an example of a lumper, whilst Pobedimova gives specific rank to four forms from the Caucasus alone which obviously belong to the aggregate. From the gardener's point of view simplicity is desirable and the species is here treated as a large and variable aggregate. The taxonomy of *C. coum* will probably not be definitely settled until its whole geographical range has been much more thoroughly explored. In the wild the various manifestations of the species probably remain true, at least within each population, but in cultivation they interbreed so freely that their individuality is usually lost.

15a. C. COUM Mill. ssp. COUM (Illustration p. 28).

(=*C. hiemale* Hildebr., *C. hyemale* Salisb., *C. orbiculatum* Mill., *C. vernale* K. Koch, and *C. vernum* Sweet).

The tuber is rounded beneath and flattened above with a thin brown skin covered with very fine dark brown hairs. The roots spring from the lower surface only. There is sometimes a short floral trunk.

The foliage is extremely variable and ranges from sombre unspotted green to the bright silver-washed green of a beautiful form (illustration p. 28: collected by E. K. Balls in Asia Minor). Between these two extremes are many intermediate forms, some of which are more handsome than others, but in general the leaves are dark to mild green, marked either with a few light green spots or with a continuous band of light silvery green outlining the central hastate pattern. In shape the leaves may be round or almost kidney-shaped. In general, the leaves of *C. coum* are rather small with a diameter of between 2.5 and 6 cm. The leaf veins usually end in small protruding whitish knobs between which the margin of the leaf is slightly concave. The upper surface is usually matt but some plants have attractive shining foliage. The under-surface may be beetroot-red, or green, or green stained with red.

The colour of the flowers varies from a not-very-clean-white, through differing shades of pink, to light or dark carmine. The rim of the corolla tube is white. Near the base of each petal there is a dark crimson or plum-purple blotch which is rounded at the top and hollowed out into two symmetrical semi-circles at the bottom so that the blotch appears to be supported on three tiny legs. This marking is characteristic of both *C. coum* ssp. *coum* and *C. coum* ssp. *caucasicum*. (*C. alpinum* and *C. parviflorum* which are closely allied species can be readily distinguished by the absence of this 'basal eye'.) The anthers are yellow and the style is not exserted. The flowers vary in size but they are generally rather dumpy in appearance with short, much reflexed petals which are almost as broad as long and slightly folded inwards. Certain forms, however, particularly those from Lebanon and Syria, have quite graceful flowers with longer, well spread petals.

C. coum is both a lowland and a montane species with an altitudinal range from sea level to 2100 m. It usually grows in woods or under bushes in damp places or near running water. Along the Black Sea coast of Turkey where the species occurs in abundance the plants can be found at sea level growing in coastal sands together with bracken, *Helleborus orientalis* and *Galanthus rizehensis*. Further inland it is a common plant of hazel coppices flowering in profusion with *Primula vulgaris* var. *sibthorpii*. At the higher altitudes it is sometimes found on rock faces. In its native lands the plant flowers after the passing of winter, often with the melting of the snow, but in cultivation the first flowers may open in December—continuing into the spring. Although remarkably hardy, both flowers and foliage are sometimes damaged by severe frost and cold winds and for this reason outdoor plants are best given a position in the garden where they will get protection from surrounding vegetation.

15b. CYCLAMEN COUM Mill. ssp. CAUCASICUM (K. Koch) O. Schwarz (= *C. elegans* Boiss. & Buhse; *C. caucasicum* Willk. ex. Boiss., *C. ibericum* Stev. ex Boiss. and probably *C. abchasicum* (Medw.) Kolak ex. Pobed.)

This is a fine plant from the Caucasus and the wooded areas of N. Iran. It is generally distinguishable by its heart-shaped leaves, relatively laıge flowers, often with graceful sub-acute or acute corolla lobes, and by its pink or pale purple 'eye' (ssp. *coum* always has a white 'eye'). If these distinctions were constantly maintained, *C. coum* ssp. *caucasicum* might merit specific rank, but there is extensive intergradation with typical *C. coum* in some areas of eastern Turkey. Examination of a representative range of specimens suggests that ssp. *caucasicum* cannot be regarded as more than a horticulturally desirable but otherwise ill-defined geographical race of the typical species.

16. C. ALPINUM Sprenger (P.D. 25579 is illustrated on p. 29).
Colour plate *Bot. Mag.* N.S. t. 437

This interesting plant was first briefly mentioned by Sprenger in *Gartenflora* 1892 p. 526, and subsequently described in detail and figured by Hildebrand. Hildebrand evidently considered it a separate species having affinities with both *C. cilicium* and *C. coum* (which he called *C. ibericum*), but the plant disappeared from cultivation and Hildebrand's specimens were destroyed, so that accurate identification of *C. alpinum* became impossible. It has been assumed, perhaps erroneously, that the material collected by Davis and Polunin in 1956, and introduced by them into cultivation, is conspecific with *C. alpinum* Sprenger; for the present this identification has been followed, though for a number of reasons, it may yet be found advisable to abandon the name *C. alpinum* and give a new name to the Davis and Polunin plant.

It was found in S.W. Turkey, growing in *Pinus nigra* ssp. *pallasiana* forest at about 1200 m. A photograph taken in the wild will be found in *R.H.S. Journal LXXXII* (April 1957), fig. 48. In shape the leaf is rounded at the base, becoming ovate towards the apex, which is somewhat truncate. The basal lobes do not overlap, and the leaf veins end in small whitish projecting points. The leaf measures up to 5 cm. in length by almost as much in breadth at its widest point. The hastate pattern is very shiny and is surrounded by a ring of lighter blotches. There is a dark edge to the leaf, broken by a few light spots, and the undersurface is deep shining red. In appearance the leaf resembles *cilicium* rather than *coum*.

The flower has a span of 4 cm., and the broadly ovate, blunt pointed, slightly dentate petals are 12 mm. long by almost as much broad. They show no folding but are twisted through 90 degrees near the base and stand out almost horizontally ('propeller like'), making the flower appear very large. The colour is rose-pink to deep carmine pink, becoming darker at the mouth. At the base of each petal is a solid, semi-circular, very dark crimson blotch whose base is straight, not three pointed as in other members of the *C. coum* aggregate, i.e. without a 'basal eye'. The flower smells strongly of heather honey like that of *cilicium*.

The main points of difference from *coum* are:—

1. The shape and marking of the foliage,
2. the horizontal poise of the petals which are not folded and only slightly reflexed,
3. the shape of the basal blotch,
4. the scent, which is absent from forms of *C. coum*.

17. CYCLAMEN PARVIFLORUM Pobed. (Illustration p. 29).

A dwarf plant with small, suborbiculate, dull green leaves and inconspicuous flowers of pale lavender-pink. The tuber is small and rounded. The reflexed corolla lobes are generally less than 10 mm. long, and the basal blotches are dark purple without a paler eye. It is obviously allied to *C. coum*, and is regarded by some authorities as being a sub-species of *coum*.

It was first collected in 1914 in the mountains about Artvin in north-east Turkey, and has since been re-found by Furse, Synge, Mathew and others in the mountains south-east of Trabzon, particularly in the upper parts of the Zigana Pass. *C. parviflorum* has been grown in gardens since 1960, but has proved a difficult plant in cultivation, probably requiring, as Brian Mathew points out (*Journ. R.H.S.* 91:387:1966), cool and moist conditions even during the growing season. It is most easily distinguished from small forms of *C. coum* by the uniform colouring of the basal blotch of the corolla; like *C. alpinum*, it lacks the pale "eye" characteristic of *C. coum*. It would appear to have little merit as a garden plant.

HYBRIDS

Cyclamen species show great reluctance to interbreed and, curiously enough, the few hybrids so far reported have occurred between species with discrepant chromosome numbers and not, as might be expected, between those belonging to the same chromosome group. The plant *C. atkinsii*, which was raised in 1852, was stated to be a cross between Miller's *coum* (30 chromosomes) and *persicum* (48 chromosomes), the latter being the pollen parent. Doubts have been cast on the authenticity of the plant, but the fact remains that the original plant was handsome and distinct with blotched crenate leaves and large white flowers with pointed petals. Modern versions of *atkinsii* merely resemble good forms of *coum* and have probably arisen by selection, or as the result of a cross between these and ssp. *coum*. Crosses between varieties of *C. coum* have long been recognised on the Continent and it is such a hybrid which is described by Hildebrand under the name *atkinsii* Hort. Hildebrand also gives an account of hybrids between *africanum* (68 chromosomes) and *hederifolium* (34 chromosomes), whilst Schwarz publishes a description of this plant under the name *hildebrandii*. Although *africanum* has twice as many chromosomes as *hederifolium*, the plants are closely related and a cross between them seems quite possible. Many of the plants which pass for *africanum* in cultivation, but are smaller and much hardier, can probably be referred to this hybrid.

Some years ago a hybrid was found at Wisley between the two closely related species, *creticum* (22 chromosomes) and *repandum* (20 chromosomes) and the resultant plants are almost intermediate between the two parents. The same hybrid has occurred fortuitously in the garden of the writer. The leaves of this plant are shaped like those of *repandum* but are smaller and greyer. The flowers, which are not as large as those of *repandum*, and have shorter styles, are either white or an unattractive shade of bluish-pink. The flowers are fertile.

CULTIVATION

In order to cultivate the various species of *Cyclamen* successfully it is desirable to study the way in which they grow in nature, otherwise they may be subjected to conditions which are unsuitable to them and which they cannot tolerate. From the foregoing account of the plants it is evident that the chief requirements are:

1. Protection from cold in winter.
2. Shading in summer.
3. A period of dormancy during the resting period accompanied by a complete or partial drying-off.
4. Shelter from wind.
5. Good drainage.

These conditions can be fulfilled by growing the plants in pots in a garden frame and, with exception of *rohlfsianum*, all the species are worth attempting, although it is obvious that the really tender kinds will succeed better and give less trouble with the aid of artificial heat.

A further advantage of growing cyclamen in a frame as against their being mixed up with other plants in an alpine house is that it facilitates collective treatment, whether it be protection from cold, shading or precautions against pests.

The frame should face south and be erected in as sheltered a position as possible. Slatted shades for summer use and a really thick covering for the winter must be provided. In addition, staples should be driven in at intervals inside the frame, either at the back and front or along the two sides, about four inches from the glass, to take extensible wires on which layers of stout paper can be laid during very cold weather. The frame is best filled with coarse sand and the pots embedded in it up to their rims and placed close together to give mutual protection. As the rims of the pots should be about seven inches from the lowest part of the glass it will usually be necessary to set the frame on a brick foundation or to dig out some of the inside soil in order to get a sufficient depth of sand in which to bed the pots.

Winter Treatment

When arranging the plants in the frame due consideration must be given to their relative hardiness and, if possible, the less hardy species should be grouped together in one portion of the frame, towards the back which is usually warmer than the front.

Cyclamen vary in their ability to withstand frost, and while some species may only lose a few of their leaves after a cold spell, others will be completely defoliated, and the really tender ones killed outright.

The species may conveniently be grouped in four columns in order of hardiness, those listed in column 1 being the hardiest, while those in column 4 are the most tender:

1. *coum*	2. *cilicium*	3. *balearicum*	4. *africanum*
purpurascens	*mirabile*	*creticum*	*persicum*
hederifolium	*pseud-ibericum*	*graecum*	*cyprium*
	parviflorum	*libanoticum*	*rohlfsianum*
	repandum		

When only a few degrees of frost are expected it will suffice to close the frame down and cover it with a light mat. For lower temperatures it is safest to lay thick paper on the wires above the tender species and cover the whole frame with some really thick material which should be large enough to hang down to the ground.

When frost is very severe, the tender species should be covered by inverted flower pots, and several additional layers of paper be laid on the wires and extra covers on the outside of the frame. While the temperature remains near freezing point during the day, all covers can be safely left in place and the frame kept shut. On warmer days a little air may be admitted, but the covers need not be removed. During prolonged cold spells the plants should be inspected at intervals, any mouldy leaves or flowers be removed and the papers dried. Once the frost is over the plants must not be immediately exposed to the air but be allowed to thaw out slowly while the covers are gradually removed. This is extremely important since it enables the leaves to regain their turgidity in a moist atmosphere before being exposed to the full light. The foliage of cyclamen suffers far less from the effects of frost when surrounded by a damp atmosphere rather than by a dry one. Therefore, if plants be efficiently covered at the onset of low temperature, much frost damage can be averted.

If, in emergency, plants have to be brought into a dwelling house they will take no harm in a room warmed by a coal fire or by electricity, provided they are placed from time to time in a box covered by a piece of glass and allowed to remain there for several hours. This will create a damp atmosphere round them and help to counteract the injurious effect of dry heat. In no circumstances must cyclamen be exposed to coal gas which is rapidly fatal to them.

Spring Treatment

As the days grow warmer, the plants in the frame will need plenty of air, but if the wind is strong enough to blow the leaves about the light should be kept on but propped well open. Most cyclamen, particularly when grown in pots, dislike having their foliage exposed to hot sunlight. Quite a short exposure will cause the leaves to become limp and drooping, and to species such as *repandum*, with very thin foliage, sunlight may cause actual damage. Those best able to endure the sun are *africanum*, *graecum* and *persicum*, but even here, pot grown plants, especially when under glass, are better given the shelter of a slat. In sunny weather, therefore, the frame must be slatted early in the day before the sun becomes hot enough to cause flagging. Cyclamen enjoy a rather moist atmosphere, and their foliage will be healthier and their flowers larger if the surrounding air is not too dry. For this reason, the frame is best kept half closed on hot days and if more moisture seems to be needed it can be completely closed for a time. On chilly evenings in early spring and late autumn it is advisable to shut down the frame for the night.

Summer Treatment

After their annual growth has finished cyclamen undergo a period of rest during which their tubers ripen and mature. The successful cultivation of these plants depends very largely on giving each species the kind of ripening it demands. From the point of view of the cultivator the species can be divided into two classes:—

1. Those that require a complete drying off during the summer with as much warmth as possible, and
2. Those that need a short resting period under cooler conditions and with just enough moisture to prevent the compost from becoming dust dry.

To class 1. belong the following species: *africanum*, *cyprium*, *graecum*, *libanoticum*, *persicum* and *rohlfsianum*.

To class 2. belong *alpinum*, *balearicum*, *cilicium*, *coum*, *creticum*, *hederifolium*, *pseud-ibericum* and *repandum*. *C. purpurascens*, being evergreen in cultivation, requires different treatment and should not be dried off, and the same probably applies to *parviflorum*.

As soon as the foliage has died down, the plants belonging to the first class must be removed to a summer frame, or large box, filled with sand, and placed in full sun. The pots should be sunk in the sand and covered with a light, propped up to admit a little air but no rain. Here they must remain without any water at all until new growth is visible in the autumn. Possessors of a greenhouse will find this an even better place for ripening the tubers. After growth has again started, a little water should be given and the plants be returned to their permanent quarters.

The species belonging to class 2. can be left in the main frame, but it will facilitate general maintenance if, as each plant dies down, it is removed to one side of the frame which can be kept closed. After the disappearance of all foliage the plants need no direct water for six or eight weeks, but the surrounding sand must be kept moist. By the middle of July most of the species will be making new growth below ground and will need some water, but it must be given sparingly at first, particularly to *cilicium*, *pseud-ibericum* and the spring flowering species.

Occasionally tubers will fail to produce growth at the appointed season but remain persistently dormant for a year or longer. This occurs most frequently in the case of *persicum*, which may take a long rest after a lavish production of flowers and seed. The phenomenon has also been observed in other species but need cause no alarm provided the tuber in question remains firm. It should then be kept just moist during the normal growing period and rested in summer in the manner suitable for that particular species. If the tuber is actually dead it will soon betray the fact by becoming rotten.

Compost and Potting

Cyclamen, in nature, grow in leaf mould, in soil containing large stones, in screes, on sloping ground and sometimes in banks or rock clefts in a vertical position. In cultivation, therefore, they demand an open porous mixture and good drainage. A close, sticky compost inhibits the growth of the roots and often causes the tuber to rot. A suitable mixture for most species consists of one part each of fibrous loam (from which all the dust has been removed), leaf mould and coarse sand, with the addition of rough mortar rubble or small stones. For woodland species, one part of peat may be added.

The tubers should be just covered by the compost and set rather low in the pot to allow for the addition of a one inch topping of very rough leaf mould or small stones. This does not retain moisture and its object is to check the evaporation from the compost and to keep the top of the tuber dry. As it is easily shaken off it facilitates the inspection of the tuber without undue disturbance.

Cyclamen must not be overpotted and are best kept in the same pot for a number of years. An occasional top dressing of coarse leaf mould and a dusting of a suitably balanced fertiliser will prevent the plants from becoming starved.

Repotting

This should not be done until absolutely necessary, for not only do most cyclamen resent disturbance but they are inclined to flower more freely when pot bound. For the majority of species the correct time to undertake repotting is early July. The exceptions are *rohlfsianum* which is best left until the new growth appears, and the evergreen *purpurascens* which suffers least if moved in March before the new season's growth has started. If there are many plants to be repotted, the autumn flowering species must be moved first.

Watering

Cyclamen dislike a water-logged compost and should be kept slightly on the dry side. During warm weather they can safely be watered from above and, in fact, will benefit from being exposed to rain. On cold, damp days, when the leaves will be slow to dry, the plants are best watered from below by standing the pots for a short time in about two inches of water. In the case of *hederifolium* and *purpurascens*, whose roots emerge from the sides and top of the tuber, care must be taken that the water reaches the surface of the compost.

PESTS

1. VINE WEEVIL. The small, C-shaped, fat white grubs of the vine weevil feed on roots and corms and are about ½ inch long when fully grown. A dozen or more grubs may develop in a single small pot, and when this happens severe damage results and the plant may suddenly collapse and die. This pest also attacks cyclamen and other plants growing in the open.

The adult weevils are dark in colour and nocturnal in habit and are therefore seldom noticed. They reproduce parthenogentically and male vine weevils are virtually unknown. Each female can lay up to about 1,000 eggs in the soil over a period of 2—3 months, and a single weevil can therefore start quite a serious infestation.

Adult vine weevils hide in leaf litter and other plant debris during the day and infestations may be checked by tidying up accumulations of leaves and areas of coarse vegetation. In addition, BHC dust may be applied to any possible hiding places and may also be used around pots and on the surface of the soil to deter egg-laying. Such treatments are mainly preventative.

If the pest is well-established before it is noticed, it may prove difficult to kill the larvae in the soil, but drenching with a spray-strength solution of BHC or malathion should kill young grubs. Any grubs seen when re-potting or planting should be picked off and destroyed immediately. Some protection is given by incorporating BHC in potting composts, but this could have adverse effects on the growth of plants, especially if recommended dosage rates are exceeded.

2. CYCLAMEN MITE. This is a serious pest of pot cyclamen grown in heated glasshouses, but may also affect plants in unheated glasshouses and frames and in the open. The mites feed and breed in the young leaf and flower buds and, although they are very small, they can sometimes be seen by opening up young buds and examining them with a hand-lens. A single bud may contain a hundred or so eggs, immature stages and adults, and such buds are usually seriously affected by the feeding activities of the mites. This results in distorted flowers and leaves and infested plants may look very unsightly.

Control is not easy as there are no really effective chemicals available to amateurs. Any distorted buds should be picked off and destroyed immediately, before the mites can spread to other plants. In addition, fine sulphur dust or dicofol sprays may be applied to protect healthy plants from invasion. Once the mites are inside the buds, these chemicals will have little effect on them.

3. APHIDS. The mottled arum aphid and various other species may attack young leaves and buds at almost any time of year. Leaves and flowers are distorted and discoloured and the growth of young plants may be severely stunted. Serious damage can be prevented by spraying thoroughly with a good contact spray, such as

malathion, BHC or nicotine, or with a systemic insecticide, such as dimethoate or formothion, as soon as any aphids are noticed.

Colonies of grey or white root aphids may develop on roots and corms and severe infestations will check growth. Detection of these pests is not always easy as they live below the surface of the soil. They should be looked for when lifting, planting or repotting plants and, if they are found, the root system may be treated with a spray-strength solution of malathion, nicotine or diazinon.

4. GLASSHOUSE RED SPIDER MITE. This pest is most likely to be troublesome on plants grown under glass, but will also occur out of doors during the late spring and summer. The first sign of attack is usually the appearance of small discoloured areas on the upper surfaces of the leaves. Examination of the undersides of affected leaves, preferably with a magnifying glass or hand-lens, will reveal colonies of the very small green or yellow mites (they are usually red only when they are starved or hibernating).

Chemical control is not easy but regular and thorough spraying with malathion, dimethoate or dicofol may reduce populations. Biological control based on introducing *Phytoseiulus persimilis*, a species of mite that preys on red spider mites, is a possible alternative, especially in glasshouses and frames.

5. GLASSHOUSE WHITEFLY. This pest is also most troublesome under glass. The adult whiteflies look like miniature pure white moths and are therefore quite conspicuous, but the larvae and pupae, which are usually attached to the undersides of the leaves, are relatively inconspicuous. Infested plants become sticky and the leaves are later contaminated with black, sooty moulds which grow on the upper surface. The insecticides recommended for use against aphids will also control whitefly, but these only kill the adults so repeat applications at weekly intervals may be necessary if established infestations are to be checked.

6. MICE. Long-tailed field mice will damage cyclamen corms in store and after planting, and the house-mouse may also eat stored corms at times. The main danger period in garden sheds, glasshouses and frames is in the late autumn and early winter, when the field mice move inside as outside temperatures fall. Intensive trapping at this time should limit the amount of damage done. There is a second danger period in the spring when cyclamen growing out of doors are attacked. Traps may be set in the open, but should be placed under tiles or panes of glass so that birds and domestic animals cannot get to them.

In addition to trapping, chemical repellents may be used to try to prevent damage, but, although various repellents are available, none of them seems to be entirely satisfactory.

DISEASES

1. GREY MOULD. This disease, which is also known as Botrytis rot, causes a spotting of the flower petals and rotting of the leaves and leaf stalks, particularly at the crown. A greyish mass of fungal spores develops on the affected tissues, with the exception of the flowers which do not normally rot.

The fungus is encourgaged by humid atmospheres, and where cyclamen are kept in a greenhouse, the disease can usually be controlled by careful ventilation. Stems of unfertilized flowers, fallen corollas and dead and dying leaves should be removed. If the trouble persists, the plants should be sprayed with captan or thiram and greenhouses can be fumigated with tecnazene (TCNB) smokes. Where the foliage is dense, dry Bordeaux powder could be shaken among the leaf stems as a precautionary measure, but this will look unsightly.

2. BLACK ROOT ROT. Cyclamen grown in unsterilized compost are susceptible to this trouble which is caused by a soil-borne fungus. The roots become black and rotten, with the result that the leaves become discoloured and wilt, and the corm can be easily lifted.

This disease can only be confirmed by microscopic examination of diseased roots. If it is suspected, however, that plants have been affected by this disease, the corms should be lifted, the roots should be washed and any which are dead or bear black patches should be cut off and burnt. The plants should be repotted in sterile compost and then watered two or three times at three-weekly intervals with a solution of 1 oz. of captan in 2½ galls. of water. It would also be worth while spraying affected plants with a foliar feed to boost their vigour and encourage the development of new roots.

3. PHYSIOLOGICAL DISORDERS. Unsuitable cultural conditions result in various types of leaf discolorations such as yellowing or brown blotching. The affected tissues may then be entered by secondary bacteria which will cause a soft and slimy rot.

These troubles can be avoided by paying careful attention to watering etc. If the leaves do become discoloured, they should be removed before they become slimy. Should a soft rot develop, however, the crown should be dusted with dry Bordeaux powder, once the diseased tissues have been removed.

SEED SOWING

The seed of most species ripens in late summer and maturity is heralded by a slackening of the coils of all those species whose peduncles roll up. The uncoiled peduncles of *persicum*, however, remain rigid until the capsules have opened. As soon as the capsule starts to split, the seeds should be removed and sown while they are plump and still covered with the mucilage in which they were embedded. They will then germinate in a few weeks or months according to the time of year at which the parent plant normally produces its foliage. Thus, seedlings of *hederifolium* can be expected in September or October, those of *coum* in early winter, and so on. If the capsules are allowed to dry up the seed becomes withered, it may take a year or more to germinate and perhaps will not come up at all. In any case, germination of old seed is usually irregular, and the young plants often experience great difficulty in emerging from the hard seed coat. Where this occurs it is helpful to wind a small pad of cotton wool round the ends of two match sticks, and to fix them against the sides of the seed so that it is sandwiched between the wads of cotton wool which must be kept moist. This will usually soften the coat sufficiently for it to be shed by the developing leaf. Freezing will sometimes aid the germination of old seed, whilst soaking it in water for 48 hours may be tried with advantage. In cases where seed has to remain unsown for several months, it should be left in the capsule and stored in a polythene bag.

Seed must be sown in a porous gritty mixture and seems to germinate better when the pans are kept in a frame rather than in the alpine house where the compost is apt to get too hot. Seedlings should be encouraged to go on growing for as long as possible and, after they have died down, be kept slightly moist, otherwise the tiny tubers may wither up. When growth starts again the young plants can be pricked off into shallow pans. Unlike mature plants, young cyclamen may safely be moved at almost any time of the year. All soil from ungerminated seed pans should be used again for other seeds. Seedling cyclamen will frequently appear belatedly and take no harm if immediately removed into a very light compost. As a rule only one seed-leaf develops, but on occasion two are seen.

CONCLUSION

The object in writing the foregoing booklet was to encourage interest in the cyclamen species and to show how a fairly representative collection can be maintained in a garden frame without the aid of expensive equipment or artificial heat. Against the housing of a collection in a frame it may be argued that the hardier species must, perforce, be unnecessarily coddled in the interests of their weaker brethren. Since, however, cyclamen are seldom killed by kindness but enjoy all the comfort they can get, no fears on this score need be entertained.

Where space allows and conditions seem suitable as many kinds as possible should be tried out of doors. Not only do cyclamen look exceptionally charming in the open, but surprising success is often achieved with the less hardy sorts, whilst there are few gardens in which some, at least, of the hardy species will not be at home.

INDEX

(*Numbers in brackets indicate sequence of species in the text*)